THE JAMES BACKHOUSE LECTURES

This is one of a series of annual lectures. The Backhouse Lectures began in 1964 when the Australia Yearly Meeting of the Religious Society of Friends was first established. The lecture is named after James Backhouse, who travelled with companion George Washington Walker throughout the Australian colonies from 1832 to 1838.

Backhouse and Walker were sent to Australia by the London Society of Friends to investigate and report to authorities on the penal system and on the treatment of aborigines, to promote the cause of temperance and to preach to the scattered settlers both bond and free. The pair travelled to all major centres and to many isolated settlements. They submitted detailed observations to local as well as British authorities and made recommendations for legislative reform. Many of the changes they initiated resulted in improvements to the health and wellbeing of convicts, Aboriginal people and the general population.

A naturalist and a botanist, James Backhouse is remembered also for his detailed accounts of native vegetation which were later published. James Backhouse was welcomed by isolated com- munities and Friends throughout the colonies. He shared with all his concern for social justice and encouraged others in their faith. A number of Quaker meetings began as a result of his visit.

Australian Friends hope that these lectures, which reflect the experiences and ongoing concerns of Friends, may offer fresh insights and be a source of inspiration.

The 2024 Backhouse Lecture *God's ways, not our ways: A dissident Quaker response to disability* was delivered by Jackie Leach Scully on Monday July 8th in Adelaide.

Jackie has been involved with Quakers for over 40 years, first in the United Kingdom, then Switzerland, again in the UK and now in Australia. She has also occupied a variety of roles, from Attender to Member to Clerk of Switzerland Yearly Meeting, and has served on more committees than she cares to remember. In 2002, she was privileged to give the annual Swarthmore Lecture to Britain Yearly Meeting, *Playing in the presence: Genetics, ethics and spirituality.*

Jackie has a scientific background as a molecular biologist, combining this with interests in philosophy and theology that eventually moved her professional focus to bioethics and medical ethics. After some years working in Switzerland and the UK, in 2019 she moved to Sydney to lead the Disability Innovation Institute at UNSW, where she is also Professor of Bioethics. Disability has shaped her family, career, personal and professional life, and her engagement with faith and spirituality. Drawing on her personal and professional experience, she looks at traditional and contemporary theological engagement with disability. She uses Quaker testimony to explore how Friends are called to respond to disability and impairment, and shares some "dissident thinking" with Australian Friends, and others, to help build a world more inclusive of all kinds of difference and diversity.

Bruce Henry,
Presiding Clerk,
Australia Yearly Meeting
July 2024

© The Religious Society of Friends (Quakers) in Australia, 2024.
Produced by The Religious Society of Friends (Quakers) in Australia Inc.
Download copies from www.quakersaustralia.info
or order from http://ipoz.biz/shop
www.youtube.com/@quakersaustralia

THE JAMES BACKHOUSE LECTURES

2002 *To Do Justly, and to Love Mercy: Learning from Quaker Service*, Mark Deasey

2003 *Respecting the Rights of Children and Young People: A New Perspective on Quaker Faith and Practice*, Helen Bayes

2004 *Growing Fruitful Friendship: A Garden Walk*, Ute Caspers

2005 *Peace is a Struggle*, David Johnson

2006 *One Heart and a Wrong Spirit: The Religious Society of Friends and Colonial Racism*, Polly O Walker

2007 *Support for Our True Selves: Nurturing the Space Where Leadings Flow*, Jenny Spinks

2008 *Faith, Hope and Doubt in Times of Uncertainty: Combining the Realms of Scientific and Spiritual Inquiry*, George Ellis

2009 *The Quaking Meeting: Transforming Our Selves, Our Meetings and the More than-human World*, Helen Gould

2010 *Finding our voice: Our truth, community and journey as Australian Young Friends*, Australian Young Friends

2011 *A demanding and uncertain adventure: Exploration of a concern for Earth restoration and how we must live to pass on to our children*, Rosemary Morrow

2012 *From the inside out: Observations on Quaker work at the United Nations*, David Atwood

2013 *A Quaker astronomer reflects: Can a scientist also be religious?* Jocelyn Bell Burnell

2014 *'Our life is love, and peace, and tenderness': Bringing children into the centre of Quaker life and worship*, Tracy Bourne

2015 *'This we can do': Quaker faith in action through the Alternatives to Violence Project*, Sally Herzfeld

2016 *Everyday prophets*, Margery Post Abbott

2017 *Reflections on the 50th anniversary of the 1967 Referendum in the context of two Aboriginal life stories*, David Carline and Cheryl Buchanan

2018 *An Encounter between Quaker Mysticism and Taoism in Everyday Life*, Cho-Nyon Kim

2019 *Animating freedom: Accompanying Indigenous struggles for self-determination*, Jason MacLeod

2020 *Seeking union with spirit: Experiences of spiritual journeys*, Fiona Gardner

2022 *Creating Hope: Working for justice in catastrophic times*, Yarrow Woodley

2023 *Quakers, the Internet, and What's Next*, Jon Watts

2024
THE **JAMES BACKHOUSE** LECTURE

God's ways, not our ways: A dissident Quaker response to disability

Jackie Leach Scully

'Disability is a dramatic reminder that God's ways are not our ways.'[1]

---

[1] Jennie Weiss Block, *Copious Hosting: A Theology of Access for People with Disabilities*. Continuum, 2002, p. 91.

© 2024 Religious Society of Friends (Quakers) in Australia
PO Box 4035, Carlingford North NSW 2118
secretary@quakersaustralia.info
quakersaustralia.org.au

ISBN 978-1922830-76-0 (PB); ISBN 978-1922830-77-7 (eBk)

© 2024, Jackie Leach Scully, The Religious Society of Friends (Quakers) in Australia

The moral rights of the author have been asserted.

All rights reserved. Except as permitted under the *Australian Copyright Act 1968* (for example, a fair dealing for the purposes of study, research, criticism or review), no part of this book may be reproduced, stored in a retrieval system, communicated or transmitted in any form or by any means without prior written permission.

Design & layout by:
Interactive Publications Pty Ltd, Carindale, Queensland 4152

Cover image: Emily Crockford, Studio A.

 A catalogue record for this book is available from the National Library of Australia

Contents

Acknowledgements	vi
About the Author	vii
Introduction	1
A Personal Background	5
Thinking About Disability	13
Models of Disability	17
Dissident Thinking	19
Theologies of Disability	21
Disability in the Old Testament	23
Disability in the New Testament	24
Theodicy: the Problem of Suffering	27
Disability in Contemporary Theological Thinking	30
Disability in Reproductive Technologies	35
Disability and Discernment	43
Being Embodied	45
Being Vulnerable, Dependent, and Not in Control	48
Being in Community	52
A Dissident Stance on Disability and Difference	57

Acknowledgements

I want to thank the Backhouse Lecture Committee for the invitation to give the 2024 Backhouse Lecture to Australia Yearly Meeting. Initially I was hesitant to accept, feeling that as a relative newcomer to this land it would be presumptuous of me to offer any suggestions to Australian Quakers. I'm grateful for the Committee's reassurance on that point, and for their help and encouragement through the process of writing it. Thanks also to my Friends in the Online Recognised Meeting for Worship.

After almost 30 years as a bioethicist, 37 years as a member of the Society of Friends, and over half a century as a disabled woman, it's impossible for me to name every person who (whether they realised it or not) helped shape my thinking on bioethics, religious faith and practice, and disability. But if I have to single any out, the list would necessarily include Alison Peacock, Debby Plummer, Christoph Rehmann-Sutter, Tom Shakespeare, Margaret Spufford, and Simon Woods. A particular thanks goes to Rachel Muers for answering a question on Bible history, and to Iva Strnadová for reading a draft of the text.

As always, thanks to Monica Buckland for this long-haul flight.

– June 2024, Sydney

About the Author

Jackie Leach Scully has been involved with Quakers for over 40 years, first in the United Kingdom, then Switzerland, again in the UK and now in Australia. She has occupied a variety of roles too, from Attender to Member to Clerk of Switzerland Yearly Meeting and served on more committees than she cares to remember. In 2002, she gave the annual Swarthmore Lecture to Britain Yearly Meeting, *Playing in the presence: Genetics, ethics and spirituality.*

Jackie first trained as a molecular biologist and spent several years investigating the mechanisms of cancer and neurodegenerative disorders. Later, she became interested in the ethical and legal issues raised by contemporary medicine, and studied philosophy, theology and psychology as she moved her focus to bioethics and medical ethics. At the University of Basel, she helped to establish the first interdisciplinary bioethics centre in Switzerland, and at Newcastle University in the UK was Executive Director of the Policy, Ethics and Life Sciences Research Centre. In 2019, she moved to Sydney to lead the Disability Innovation Institute at the University of New South Wales, where she is also Professor of Bioethics.

Disability has shaped her family, career, personal and professional life, and her engagement with faith and spirituality. Now making her home in Australia, she tries to follow the suggestion of Britain Yearly Meeting's Advices and Queries 27, "Live adventurously".

Introduction

This Backhouse lecture will focus on the meaning of disability and impairment for Quakers today. In doing so, it also asks a broader question: how should the Religious Society of Friends in Australia in the 21st century respond to embodied difference and variation? In what follows, I can't promise to provide an answer, but I hope to provoke thinking among Friends that will further our spiritual and practical engagement with disability and disablement.

In the first substantive section, I outline why this subject has interested me personally and professionally over the decades. After this, I look at different attempts ('models') to comprehend what disability is, where it comes from, and how best to respond to it. The third section examines some Biblical and theological resources for thinking about disability, especially for making sense of suffering and of the place of difference and disability in the created world. Next, I discuss recent advances in technologies that identify the presence of certain anomalies before birth and argue that our unprecedented ability to select against embodied difference in this way demands that society (and our Society) thinks much more deeply than before about its responses to disability. Then, in the penultimate section, I use the lens of our testimonies to suggest how Friends might begin the process of exploring the moral and spiritual meaning of disability in the 21st century. I'll be arguing that we need to approach this with *dissident thinking*[2], challenging easy assumptions and being 'open to new light, from whatever source it may come'.[3]

---

[2] I've chosen to use the language of dissidence rather than dissent here, because the latter is so closely tied to the history of dissenting church groups like the Quakers. To think dissidently doesn't mean to think in line with a particular church tradition. To my mind, dissidence also carries a stronger sense of action as well as opinion; thinking *and acting* differently.

[3] Britain Yearly Meeting, *Quaker Faith and Practice: Advices and Queries*. London, 1995.

Just a couple of further points, both on language, before we get stuck into it. The first is about the terminology of disability. I shall have more to say about it later, but at this point I need to explain my own choice of words. Those who are familiar with the world of disability scholarship and activism will know that language is highly contested, with even closely related countries taking different approaches. The United States, for example, prefers 'people first' language when referring to most minority groups, on the grounds that this centres the person rather than the contingent feature: so, a *person of colour* rather than *black person*. In line with this, their preferred usages for disability are *people with disabilities* or *disability, person with vision impairment, person with limited mobility*, and so on. Australia generally uses *people with disability*. In the United Kingdom, on the other hand, the preference is for *disabled person*. This is because Britain was the origin of the social model of disability (more on that later), and this phrasing reflects the view that a person is actively *disabled* by society.

Language is vital in shaping beliefs and values, and so I don't dismiss word choice as trivial. On the other hand, it's worth remembering that what ultimately matters is behaviour (as a colleague once said to me, 'I don't care what people call me as long as they treat me right'). Since I come from the UK but now live in Australia, I tend to alternate between *people with disability* and *disabled people,* and I hope any readers who find either of those terms problematic will accept my reasons for doing so.

The second is about the religious language and concepts I use in this lecture. My background is religiously mixed, but I was raised within the predominantly Christian culture of Britain in the 1960s and 1970s and using traditional God-language. There was a period in my twenties and early thirties when I rejected that language and searched for alternatives, including familiar Quaker ones such as the Spirit or the Light, and others such as Ultimate Reality, the Eternal, and so on. But, in the end, I came back to the older terminology for a number of reasons. For one thing, I realised that any words here are an attempt to conceptualise something that is essentially beyond concepts, and so are always doomed to be inadequate. Given that, I decided I might as well go with *God* since that's what I'm most used to, and at least it has the advantage of being short.

Although I would not describe myself as a Christian as that term has traditionally been understood, I have enormous respect for the intellectual and spiritual work of the Christian theological tradition, which I know better than any other. I'm comfortable using Christian language and the resources of the Bible, even if I interpret and use them rather differently than the people who taught me. But I'm aware that some readers, including non-theistic Friends, may have difficulty with this language and concepts, so here and there I've used alternatives. I hope that the verbal packaging doesn't prove a barrier to the actual content of what I'm trying to say. Readers should feel free to mentally translate the terms I use into something that makes more sense to them.

Later on, I look at some of the ways that the Bible, the main Judaeo-Christian scripture, handles disability, and I discuss how some Christian theologians approach it. I want to emphasise that this doesn't mean I think the Bible is the only source of spiritual insight or that Christianity is the one true religion. I'm just interested in past and contemporary interpretations of artefacts like the New Testament's healing narratives, because these interpretations have exerted such a profound and inescapable influence on our cultural negotiation with disability and illness. As a disabled child, when these miracles were presented to me, they carried the clear message that, from Jesus' point of view, I'd be a priority candidate for repair; that, unlike the other kids around me, I was not good enough as I was. For me, whether those miracles 'really happened' or not isn't particularly relevant. More interesting, and I think useful, is to understand how, at any point in time, different interpretations of these accounts both reflect and influence our beliefs about disability and difference.

In this lecture, when I consider how theologians of the past and present have tried to make sense of disability and difference, I don't discuss non-Christian traditions. Again, this is not because I believe other faith traditions have nothing to say to Quakers of the 21st century, and my own thinking continues to be enriched by learning from other traditions. But, just as much as the first Quakers of 17th century England, I've been injected into a particular culture, place and time that set the parameters of my experience.[4] I'm a bioethicist with an interest in the ethics of disability, not a scholar of comparative religion. I know I can't do proper justice to the thinking of other religions and must leave that for others.

---

[4] I'm grateful to Debby Plummer, teacher and priest, for giving me this analogy more than 40 years ago.

A Personal Background

Disability has been part of my life for as long as I can remember, though it has changed through time as my body and its interaction with the world has changed. In fact, my exposure to disability began the moment I arrived in the world. My mother had what we would now call a congenital birth anomaly:[5] she was born with only one fully developed arm, the other ending in a fingerless stump just below her elbow. It looked very much like the effect of the drug thalidomide that caused so many birth anomalies around the world in the early 1960s, but in fact this kind of anomaly happens spontaneously in around 1 in 70,000 births. Growing up with a one-armed mother meant that I took it as quite normal that people could vary in this way, a point I'll return to later.

Then at the age of eight I contracted meningitis, an inflammation of the meninges, the membrane around the brain. Even today, meningococcal meningitis has a mortality rate of 5-10%. It leaves around 20% of patients with lasting health consequences, including (as in my case) sensorineural deafness. My parents were grateful to have a live child at all, and so did not at first grasp the profound difference this would make to my life. Virtually every moment of every engagement with the world has to be done differently if hearing can no longer be a part of it. People with disability come to know, in a way that nondisabled people generally do not, that an impairment does not have to be severe to affect life on every front. My deafness has influenced everything from my education to the activities I enjoy, to how I arrange the furniture in a room. I was always rather introverted and bookish, and I suspect that was exacerbated by having major difficulties communicating with people. I can't say I did well at school *because* I was deaf, but it certainly meant I was more likely to be found doing my homework than going out with a gang of friends. It also

---

[5] By using the word 'anomaly' I'm indicating that a bodily form is statistically unusual (that is, anomalous), but don't intend any value judgement about whether an anomaly is disabling or not. I prefer anomaly to abnormality, which to my mind has a stronger negative connotation.

meant I never seriously considered a career involving a lot of face-to-face interactions. In subtle ways like this, personality and disability interweave.

After the age of 16, I specialised in science subjects, going to university to read biochemistry and on to a PhD in molecular biology, then about a decade as a research scientist. Although my deafness had some effect on my personality, and influenced my development into a nerd, for much of my career it had no noticeable place in my professional life. Through a meandering series of chance events and decisions, I found myself gradually moving from the laboratory bench and into the field of bioethics. This is a broad discipline that is concerned with ethical issues in medicine and other life sciences; everything from very fundamental ethical questions (like, 'what is the moral status of an embryo?') to questions of regulation and health policy (such as, 'should fertility treatment be publicly funded?'), and all points in between.

Because my life as a scientist had focused on the molecular genetics of diseases like cancer and dementia, these were the areas I initially found myself working in as a bioethicist. Later, I specialised in the ethics of new reproductive technologies such as in vitro fertilisation (IVF), surrogacy, embryo donation and so on. Gradually, from the early 2000s, my research work began to touch more closely on questions related to disability. This was almost inevitable, given the emergence of *selective* reproductive technologies. A large part of bioethics currently is about the possibilities of and limits to choice, particularly when it comes to reproduction. Today's medical technologies offer a historically unprecedented degree of choice over the number and nature of the children we have. In countries like Australia, there is overwhelming support for the idea of women's rights to bodily autonomy, and a consensus that it's acceptable for contraception and, under some circumstances, abortion to be available. We also accept that people who have difficulty conceiving can turn to assisted reproductive technologies, again within defined legal limits. These technologies offer control over *whether* and *when* we have children.

Much more ethically controversial are the technologies that enable us to select the *kind* of child we have. In practice, this typically means identifying anomalies before birth and deciding whether or not to continue with the pregnancy. Being able to select against disability and impairment raises enormous moral challenges, some of which have brought bioethicists and disability activists into confrontation. I will be saying more about using prenatal screening to select against disability later on.

Working as a disabled bioethicist, I became aware of two important things. One is that bioethics' close relationship with medicine means that it is *centrally concerned* with disability, because much of medicine is really about preventing or ameliorating conditions that are disabling. Bioethicists have been asked (or felt obliged) to evaluate the ethics of prenatal screening for disability, but also of other areas of medicine such as research into treatments and cures, the rationing of healthcare, end of life services – all of which usually involve evaluating the *quality of life* of people with a chronic illness or disability. In fact, many bioethicists today are uncomfortable with this, avoiding the term quality of life because of its close association with the discipline of health economics, which tries to quantify the experience and 'value' of a life. A better concept is provided by the idea of *flourishing*, which goes back to the philosopher Aristotle and his discussion of *eudaimonia*, or how to live a good life. A flourishing life can take many forms and is more open to subjective assessment by the person actually living the life, or those who love them.

The second point is that although bioethics is a highly differentiated and sophisticated field, its treatment of disability is often overgeneralised and unsophisticated. When writing about disability, bioethicists may occasionally refer to specific conditions like Down syndrome or spina bifida, but it's much more usual to see vague gestures towards 'disability' or 'severe handicap'. Because individual bioethicists tend not to know very much about life with impairment, they also often make assumptions about the associated quality of life, assumptions that can be flawed or outright wrong. Part of the problem is that there are still not many professional bioethicists who (a) write about disability and (b) identify as disabled; I can think of fewer than ten around the world, including myself.

Taken together, these two things meant that I found myself drawn into thinking critically about bioethics' engagement with disability. I now try to distinguish between the *bioethics of disability* – the systematic reflection about the morally correct ways to behave towards people with disability – and *disability bioethics*, by which I mean the distinct moral understandings that are generated through the experience of disability.[6] Both of these approaches are valuable in different ways, but I think only the perspective from experience gives us insight into the spirituality of disability.

There's a point in most disaster movies when a character exclaims, 'Everything's gonna be alright now!' – but the viewer knows there's still 40 minutes to go and ample time for a lot more to go horribly wrong. Something like that happened to me in 2015. I had been writing and publishing about disability bioethics for about 15 years and felt, perhaps smugly, that I had got my ideas sorted out. Unexpectedly, I fell acutely and severely ill with a viral infection that ultimately destroyed my liver, and, within a fortnight, I underwent an emergency transplant operation. Transplants are not cures, and since that time I've had a very different encounter with disabling chronic illness and impairment, and have been trying to make sense of this new experience.

This 'making sense' of disability and illness is not just as an ethicist, but as a Quaker. Coming from a family that was mixed race and hybrid in terms of culture and religion, I was both a baptised Catholic and confirmed into the Church of England, eventually finding my way to the Religious Society of Friends when I was in my late teens. I joined formally at the age of 25 and have been a member of Yearly Meetings in Britain, Switzerland, and now Australia. I can honestly say that, through all this, I have never found my scientific and my spiritual predispositions coming into conflict with each other; the most serious struggle has been trying to make theological or spiritual sense of the suffering that can result from illness and disability (and of course from many other human experiences as well). I'm an ethicist by profession and claim no great theological expertise. Many years

---

[6] Discussed in much more detail in my *Disability Bioethics: Moral Bodies, Moral Difference*, Rowman and Littlefield, 2008.

ago, I did undertake formal study in theology, but only for a short time.[7] Nevertheless, I've been fortunate to have friends and colleagues who have grappled in a serious way with these issues, as ethicists and theologians, and often against a background of personal experience of various kinds. It's been with their help that I've tried to think through questions of disability, suffering, and belief.

---

[7] Although for reasons too complex to go into here, I spent about a year having my salary paid by the Department of Theology at Basel University.

Thinking About Disability

We all have a picture in our heads of what 'disability' is. The details often depend on the contact (or lack of it) we have had with disabled people, or the presence of disability in our family or community. In fact, sheer familiarity plays a significant role in how we set the parameters of normality and abnormality. I mentioned earlier that my mother had only one arm; as a result, one-armed mothers were my normality, and I still feel no sense of *ab*normality when I see people missing one or more upper limbs.

Defining disability in words, however, is challenging. For one thing, disabilities are very diverse. Any concept of disability must somehow bring together everything from paraplegia to various genetic syndromes, mental illnesses, sensory impairments; disablement that results from chronic health problems or that follows an accident; conditions that are present from birth and those that develop later; impairments that are age-related; and disabilities that are stable or progressive, or that fluctuate over time. It also has to account for the fact that people with the exact same bodily variation can disagree on whether they are disabled at all.

The category of disability itself is historically recent.[8] Before the 19th century, impairments such as blindness, deafness, being lame or being 'mad' were considered distinct conditions and not gathered under one heading. Grouping everything together has bureaucratic and administrative advantages for providing medical or social care, but it also has the disadvantage of erasing the very obvious fact that these are distinct ways of being. In addition, there is the problem that this grouping together effectively splits people into two supposedly separate states: disabled and nondisabled.

The language available to us to talk about disability presents another problem. There is almost no neutral terminology: words like *dis*ability, *dis*order, *mal*function, *in*valid, *de*formed, or *dis*figured make

---

[8] Anne Borsay, *Disability and Social Policy in Britain since 1750: A History of Exclusion*. Palgrave, 2005.

a clear negative judgement even before we begin any discussion. It's also interesting that the word 'disability' is often used as if it describes an overall characteristic of people, directly equivalent to the labels of sex or gender, ethnicity, age, sexual orientation, and so on. But these other labels are in themselves neutral (the word 'ethnicity' is neither positive nor negative), even though subcategories within them (such as male versus female, or straight versus gay) are often ranked in discriminatory ways. Unlike these feature labels, the word disability is *already* negative. There is no overarching, neutral word that provides a concise way of saying 'the general category of characteristics, one manifestation of which is having an impairment'. When we point to a difference and call it a disability, we have already made a fundamental evaluation of it.

This matters, because language affects not only how disabled people are treated, but how they and their lives are conceptualised. To call someone a *wheelchair user* says one thing about a person, their assistive technology, and their agency; describing them as *wheelchair bound* says something very different about all of those.

And, finally, what you might call our 'cultural repertoire' of knowledge about disablement, our sense of what it is actually like to live with various kinds of disability, is very sparse. I think that this is largely to do with the fact that many people are instinctively repelled by the thought of disability. (When I first started working in this field, I was sometimes asked, 'Why do you want to spend your days with something so depressing?') As a species, we carry a collective memory of a time when becoming disabled would almost inevitably mean death. We also seem primed to notice, and often fear, even very slight differences in appearance or behaviour, and humans have a natural reluctance to engage with things we find frightening or distressing. Culturally, I think we are still shaped by these inchoate fears around abnormality and strangeness. All of this has resulted in widespread ignorance about disability – at least, until we ourselves are personally affected by it. Then where do we turn for knowledge and understanding?

Models of Disability

Disability has often been understood as a punishment for some kind of moral or religious transgression. As we'll see later in the discussion of disability in the Judaeo-Christian tradition, it can be a community or ancestral transgression as much as an individual one.

In most parts of the world, however, science and particularly medicine has taken over from religion in providing a meaning for illness and disability. Biomedical explanations of impairment and disability are often referred to as the **medical model** of disability. In a medicalised perspective, disability is a defect or deficit, as determined by reference to a standard of physical or mental structure or function. Since the 18th century, medical science has been concerned with measuring deviations from these standards. (Your doctor says 'your blood pressure is outside the normal range', or the health visitor tells you that your baby is hitting developmental milestones faster than most other kids of their age). Societal or environmental factors don't really come into it. Even when illness or disability has an environmental cause, the actual problem is seen as an individual one located firmly within a person's body. When a deviation is considered a defect or deficit, it presents a problem to be solved, and from the medical perspective the obvious way to solve the problem of disability is through a medical intervention that prevents, cures or ameliorates it.

Within the last 50 or so years, some alternative ways of thinking about disability have been developed. These are often described as **social** or **social-relational** models, or sometimes **hybrid** models, and what they have in common is the idea that disability isn't a thing in itself but emerges out of the interaction between a person's body (physical or mental) and the world in which they live.[9] Versions of social-relational models differ in exactly how they describe the interaction between body and world, and the problems they see being generated. The first social model of disability was a strongly political one, arising within British

---

[9] The organisation People With Disability Australia describes the social model here: https://pwd.org.au/resources/models-of-disability/. There is a deeper discussion, and some criticism of social model approaches, in chapter 2 of Tom Shakespeare's *Disability Rights and Wrongs Revisited* (Routledge, 2014).

disability activism in the 1970s and 1980s before spreading to other parts of the world. It was prompted by disabled people's frustration with a purely medical perspective that simply failed to take into account the social and economic features that exacerbate or even cause disability.

A key feature of the original British social model is that it tries to separate the *bodily anomaly* (which it calls the impairment) from the *disabling impact* of the anomaly (which, confusingly, is called the disability). To make this clearer, take the example of a wheelchair user. This person has a biological impairment, in that their legs don't work like other people's do. But the extent to which they are actually disabled will depend very much on features of their environment. It could be that one way to tackle the impairment is through a medical intervention, but it could also be that it would be less disabling if ramps were available and public transport made accessible.[10]

It's a statement of the obvious to say that we try to avoid or ameliorate disability because we believe it is associated with suffering: disabled people often experience physical or mental pain, disadvantage in education or work, social exclusion and so on. Critics of the medical approach to disability say that it is good at identifying impairment, but that it's mistaken about the real source of much of the suffering associated with disability. The medical approach assumes that deviating from a biological norm inevitably leads straight to disablement, and therefore suffering. Social-relational approaches, by contrast, say that the suffering of disability comes as much from mismatches in the interaction between the body and the world.

Critics of social-relational approaches, for their part, are sceptical about the idea that the suffering of disability has little or nothing to do with bodily variation. The extreme version of the social model seems to suggest that if societies were sufficiently accommodating then, although impairments would remain, the barriers that create disablement (and the associated suffering) would simply disappear. This might be true for some impairments, but for many others the argument is just not convincing. My life would be a lot easier if accommodations for hearing

---

[10] Although described as environmental features, it should also be recognised that the absence of ramps and accessible public transport is ultimately a social and political choice.

impairment were more routinely available, but no amount of social or environmental reorganisation will help with the tinnitus I sometimes get as much as the medication I can take.

The early versions of the social model also tended to focus solely on the material aspects of disability, mostly on the discrimination that prevents disabled people from entering the workforce. They were less interested in the role that language can play in shaping people's attitudes to disability, or in the emotional aspects, like loneliness or depression, that are often the most painful parts of the experience of disability. By (over-) emphasising the political and economic side, so the critics say, the social model neglects the real complexity and diversity of the relationships that people may have with their disabilities. More recent approaches have, thankfully, tried to be more comprehensive, and leave space for the whole diversity of disabled lives, including the spirituality of disability.

Dissident Thinking

People often ask what relevance all this theoretical stuff has to the everyday problems of people with disability. But ideas about the causes and meanings of disability have serious practical implications; for one thing, they determine what are considered to be the appropriate ways to respond to it. Medicalised approaches see disability as caused purely by physiological deviation, having no intrinsic meaning, and so the best response is to eradicate disability as much as possible. Social-relational accounts locate the problem of disability in society, or in the relationship between the individual and society, in a way that reveals how oppressive and discriminatory the world is. From this perspective, the best response is to ensure that social structures and organisations become more accommodating of variant or nonstandard bodies. Bioethics, the field that examines the ethical and social implications of medical advances, has tended to follow the medical viewpoint. Disability bioethics, as I defined it earlier, tries to take a broader view, starting from the lived experience of disabled people, bringing in social justice arguments, and considering how best to meet a variety of care and dependency needs in the community. But none of these perspectives pays much attention to theological or spiritual views on disability.

In fact, I want to suggest that no single model or approach can explain a phenomenon like disability in its entirety. Instead, we should be aware that different approaches will be more or less appropriate or useful in particular contexts, The medical model will get me a diagnosis and access to treatment, and a diagnostic label may also be the gateway to additional social care, educational supports, and so on. A social-relational approach will help me argue for adaptations in my workplace, and may also shore up my sense of self-worth, if it helps me to feel that it's society, not my body, that is to blame for my disablement. Dissident thinking stands back from all of these, seeing their explanatory value but also their flaws; understanding that the person with disability also, always, holds 'that of God', and so is always *more than* any explanatory framework.

Theologies of Disability

In this part I want to look at attempts to comprehend disability and chronic illness made by past and contemporary theology. For many people, faith traditions are irrelevant to contemporary issues, and it is true that in the broadly secular west the church has nothing like the social power and influence it once did. Yet our entire culture is still shaped by its Judaeo-Christian origins, along with more recent influences, and the roots of the Religious Society of Friends are firmly in Christian belief, however far we may have diverged from that since the 17th century.

Early religious thinking necessarily turned to the canonical scriptures for guidance on a topic. But anyone looking to the scriptures for a straightforward account of the meaning of disability is doomed either to disappointment or (perhaps just as bad) to find exactly what they expect to see. As an overarching concept of disability did not exist in antiquity, we cannot expect to find its equivalent in any scriptural texts, within or outside the Judaeo-Christian tradition. Neither the Old nor the New Testament texts describe health conditions in ways that allow them to be readily identified today; moreover, they were reported and interpreted in a very different conceptual landscape. Yet while the Biblical accounts of disability are diverse, inconsistent, sometimes contradictory, they do suggest that the meaning of disability was more nuanced at various points in antiquity than many later users of the scriptures have acknowledged.

Disability in the Old Testament

Throughout the Old Testament, there is an overriding message that disease and its consequences (including disability) are a demonstration of what happens when the covenant people are not faithful. This in itself is significantly more complex an idea than just being a punishment for individual people doing something wrong. Evidence from archaeological and other sources show that, in the societies of the ancient Middle East, people with disability were considered defective, not quite as worthy as other people, and it was for this reason that those who were blind or lame (or, interestingly, infertile or 'barren') were not permitted to serve as priests, because only people without physical flaw could approach God so closely. Nevertheless, people with disabilities still counted as *people*: they were not completely invisible or treated solely as objects of charity.

Both Old and New Testaments contain accounts that feature people who are both central to the plot and clearly what we would nowadays call disabled. Moses, famously, had a speech impediment; Jacob was permanently injured during his struggle with the unnamed angel. Sometimes the impairment has a role to play in the narrative (e.g. Moses' trust in God enables him to overcome his stutter), but in others it is just a peripheral attribute of the character, with no apparent message or meaning. In one example, all the more compelling because it is so mundane, there is a passing mention that Mephibosheth, son of Jonathan, lost the use of his legs after being dropped by a nurse at the age of five (2 Samuel 4:4). In some stories, disability is a direct result of divine action: God strikes blind the people attacking Lot's guests (Genesis 19:11), while Ezekiel is deliberately rendered mute until he becomes God's mouthpiece (Ezek 3:26–27). Sickness is sometimes described as the result of separation from God, but there are also instances where disability is represented purely as a natural phenomenon: for instance, sight failing with age is mentioned in the stories of Eli (1 Sam 3:2) and Ahijah (1 Kings 14:4). In other words, in the Old Testament there is no single consistent rationale or meaning attributed to disability – hardly surprising, given the time span and cultural changes that these texts are thought to cover.[11]

Disability in the New Testament

The treatment of disability in the New Testament is dominated by the stories of the healing miracles. The events described are conventionally interpreted as a vehicle for Jesus to demonstrate his divine nature by

---

[11] Dating the Bible is notoriously difficult, but it is generally accepted that the various authors of the canonical texts were writing over a period of roughly 900-1000 years.

curing various illnesses and disabilities, or even bringing the dead back to life, just through his touch or voice. A closer look, however, suggests that these accounts – and the cures themselves – are not as straightforward as they appear. To understand what is going on, it is important to keep in mind that, at the time, the taken-for-granted link between suffering and sin meant that people experiencing disability or illness often became social outcasts. Against this background, it is striking how often Jesus is shown to be interacting with these pariahs in a direct and unconstrained way.

In fact, relatively few of the accounts are simply about the removal of a physical or mental condition, and many include other details that radically modify the point of the story. In the Gospel of John, for example, Jesus encounters a man described as having been blind from birth.[12] Following the understanding of the time, the disciples ask Jesus whether it was the man himself, or his parents, who had caused the blindness through their sin. Jesus categorically rejects both options, saying instead that the disability is there so that 'the works of God should be shown in him'. Because Jesus then goes on to restore the man's sight, these 'works' are usually taken to be the miraculous healing, but some commentators have interpreted it as referring to the blindness itself, raising very challenging questions about the point at which the glorious diversity of creation becomes disability.

The story of a paralysed man lowered down through the roof for healing can be interpreted as being about loyalty and solidarity rather than the problem of a disabled body.[13] The text tells us that it is the behaviour of the man's friends that prompts Jesus to say out loud, 'Your sins are forgiven', and it is only when criticised for doing so that he adds (with more than a hint of exasperation), 'Take up your bed and walk.' 'Your sins are forgiven' is as much an observation as a command: instead of absolving the man's sins with this statement, Jesus might equally well be seen as pointing to the obvious affection between all of these men as proof that this man's paralysis has not cut him off from his community.

---

[12] John 9:1–12.
[13] Luke 5:17–26.

My final example, although not usually counted as a healing miracle, illustrates a response to disability as difference. In yet another well-known story,[14] Jesus invites himself to dinner with Zacchaeus, the tax collector, a man whose job meant that he was considered a collaborator with the occupying Roman forces. Zacchaeus is also described as being so short he has to shin up a tree to see over the crowd, and for contextual reasons some current scholarship suggest that Zacchaeus may have been more than just a bit under average height, but an actual dwarf or person with restricted growth. If so, then Jesus is doing something doubly radical: choosing the company of someone who is not only despised because of his occupation but also (as commonly happened with dwarfs) likely to be ridiculed for his disability. And, in this account, there is no physical healing. What Jesus is most concerned with isn't Zacchaeus' physicality but the oppressive hierarchies and rules within which he is entangled.

I have discussed the healing narratives in some detail because the interpretation that they are all about the necessity of a cure for people to be whole has so deeply influenced our culture's attitude towards disability and illness. It makes a difference if we can understand them from a dissident perspective, as challenging the contemporary meaning of disability rather than (as we would today assume) being about changing what people can and can't do so they can better fit in with the norm. The restoration that happens is not about people's function but about their place within their community. Almost all of the healing miracles involve some kind of touch, again not because magic flows from Jesus' hand, but as a material sign of connection and solidarity with those cast out. It's in line with this understanding that, when Jesus talks about the Kingdom of God in what is known as the parable of the Great Banquet, 'the lame and the blind' who are invited to the meal aren't there to be healed but to be welcomed exactly as they are.[15]

---

[14] Luke 19:1–10.
[15] Luke 14:16–24.

Theodicy: the Problem of Suffering

Let's now turn from these scriptural scraps of evidence to look at what later theologians did with them to reflect on the meaning of disability. The Christian theological tradition has considered disability and chronic illness through two main lenses: how disability relates to suffering and the problem of evil, and the place of the disabled person in the created world.

In the philosophy of religion, **theodicy** is the term for attempts to solve the puzzle of why, if God is omnipotent and benevolent, evil exists at all. Suffering is taken as a manifestation of evil, and since for most theologians through the centuries it has seemed obvious that disability or illness necessarily involves suffering, they have largely conflated the two and addressed the suffering of disability as one example of the problem of evil in the world. For this to work, disability must always and inevitably be a bad thing. Yet, just as disability scholars and activists are divided on the fundamental cause of disability, it is also the case that not all people with disability see their impairments as unambiguously harmful or bad.

Although not specifically addressing disability, the Old Testament story of Job is considered the most extended examination of the meaning of suffering within the Hebrew scriptures.[16] Superficially, the message of the story is the simple and not terribly appealing one that suffering is how God tests human faith. Read as a whole, though, the text is more complex. It does not lay out a lesson but offers a debate, through different voices, in the course of which several points of view are put forward, working through various possible reasons for human misery. In the end, there is no firm conclusion about the reason for any individual's suffering, or for human suffering in general: why Job and his family should have experienced so many undeserved horrible things remains, unsatisfyingly to many, a mystery.

But the ambiguity of the debate is another rebuttal of the idea that disability is straightforwardly a divine punishment for human sin. About the only thing the story *does* make clear is that there is

---

[16] Some scholars also consider the Book of Job to be one of the oldest texts.

no one right way to respond to suffering. Job is initially accepting of whatever happens to him or his family, but later becomes angry and bitter, demanding that God answer with some kind of explanation. The important point seems to be that, throughout everything that happens, and whether his response is acceptance or raging defiance, Job remains in relationship with God.

There have been many later attempts to answer the question of why people suffer in a world supposedly created and maintained by a benevolent entity. Possibly the most familiar, and certainly the most influential, comes from the writing of St Augustine of Hippo, around 418 CE, about original sin. Augustine argues that God created the world perfectly (so no evil, disability, death or suffering), but at the same time also gave human beings free will, therefore leaving open the possibility of their going astray. According to this theory, it's because of the original sin of Adam and Eve, misusing their free will to disobey God, that evil and suffering became part of human life. Their deviation from God's plan means that all human life is now corrupted, and experiences of suffering are the inevitable result.

This account has had enormous influence, effectively shaping the western church's policy on suffering for centuries. Nevertheless, there are some fairly obvious objections to Augustine's theodicy that make it less acceptable to modern thinking. One is a basic problem of unfairness. Why should people continue to suffer now for something that the Biblical first humans, at best living a long time ago and at worst purely fictional, are supposed to have done? If God really is omniscient, then the consequences of giving humans free will were as predictable as handing a toddler an Easter egg on Saturday night and saying, 'Don't open it till tomorrow.' In that light, the suffering of all subsequent generations seems even more grossly unfair.

Moreover, the free will account of the origin of evil works best for what are called moral evils, the ones that result from humans continuing to exercise their free will by choosing badly – murder, lying, cruelty and so on. Natural evils like earthquakes, and also diseases or genetic conditions that cause disability, are more a reflection of the way the world's features are organised and, as sceptics have argued, surely an omnipotent God could have produced a world without such obvious design flaws.

An alternative influential theodicy was developed by Irenaeus, a less well-known early Church figure living in the 2nd century CE. His idea was that evil is necessary for developmental reasons, and he tried to illustrate this by distinguishing between the metaphors of the image and the likeness of God. Humans, he thought, were created in the *image* of God, meaning they have the potential to achieve spiritual and moral perfection but haven't got there yet; once they do, they will be living the *likeness* of God. But to get from the original state of potential to one of achievement, all people have to go through a kind of testing ground. To learn the lessons they must, humans not only have free will (so they can make mistakes if need be), they are also at an *epistemic distance* from God – that is, we can't have God's insight into what the right thing to do is because that would make it too easy, like looking at the back of the book for the answers to the test. We have to acquire that knowledge through (among other things) being separated from God and experiencing suffering. Evil exists as a necessary tool to develop human spiritual and moral capacity. Essentially, our experiences of pain, disadvantage, disillusionment and so on are lessons to help us on the road to becoming the likeness of God.

In some ways, this proposal – the world as a 'vale of soul-making' as the theologian John Hick put it[17] – sits more easily with us in the 21st century than the Augustinian idea. We are far less comfortable with the language of sin, and happier to think of challenges as character building, or that people who have had to struggle have learnt useful lessons. Still, the Irenaean explanation for evil and suffering also has its critics. An immediate problem is that many people go through suffering that is disproportionate to any conceivable lesson that might be learnt, or in circumstances where learning seems impossible: when the suffering is fatal, for instance, or when it happens to infants or children. As with the Augustinian notion of the entirety of humankind being forever punished for Adam's original sin, this seems utterly incompatible with a reasonably fair God, let alone a loving one. In the

---

[17] John Hick, *Evil and the God of Love*. Palgrave, 2010 (1972)

context of disability, many people who experience isolation, frustration or discrimination because of their impairment would vehemently reject the idea that God made them disabled, or allowed them to be disabled, just so that they could learn how to become more virtuous or, even worse, to be an inspiration to others.

Disability in Contemporary Theological Thinking

After centuries of indifference, over the last couple of decades there has been an upsurge of interest in disability among contemporary theologians. These writers differ from their predecessors in that many (though not all) are themselves disabled and have grappled personally with making religious sense of their experience. They have also been influenced by social-relational models of disability, and often reject the assertion that disability is straight up and down suffering. Perhaps because of that, they have been less concerned with theodicy and the spiritual mechanics of suffering than with how disability, and people with disabilities, fit into our understanding of a world created and sustained by God.

Western culture has a long history of disabled people being seen as just *less than* 'normal' people: less competent, less autonomous, less adult, less valuable. There are philosophers who have effectively denied them full 'personhood', and their legal status (to inherit, for example) has sometimes been challenged[18], particularly in the case of intellectual disability or where difficulties in communication have been equated with lack of cognitive ability. In line with this, the traditional theological view sees disability as a spiritual as well as physical or mental flaw, a falling short of God's plan for humanity. Contemporary disability theologians reject this and are trying to disentangle disability from ideas of sin and moral failure, instead building an alternative vision in which disability is central to the understanding of God and God's creation. The question for them is: can disability have a meaning that

---

[18] See Eva Feder Kittay, *Learning From My Daughter: The Value and Care of Disabled Minds*. Oxford University Press, 2019. Kittay is an American philosopher whose work has been influenced by her daughter, Sesha, who has profound developmental disabilities, and, in this book, she counters some of the dehumanising philosophical arguments I have mentioned.

is more than a lesson on the wages of sin or the need to grow as moral beings?

It's interesting that these contemporary writers often go back to the idea of the image of God, the *imago Dei*. Just as traditional philosophy has tried to define personhood through a list of things persons can do, traditional theology has often tied the possibility of living as the image of God to ideals of human form and function that inevitably exclude many disabled people, especially those who are highly dependent or have intellectual impairments. Not before time, today's disability theology questions whether qualities like autonomy or the capacity to reason really are the cornerstones for an individual to be considered to be living as the image of God. Hans Reinders, who has a son with intellectual disability, is one of many who argue that the metaphor of the *imago Dei* applies to all human beings equally, conferring the same worth and dignity on everyone irrespective of disability.[19]

About 30 years ago, Nancy Eiesland published a groundbreaking book[20] in which she pointed out that the core iconography of Christianity – its branding, if you like – is a body that is visibly imperfect. The bodily wounds of Christ are visible at his death on the cross and on his corpse after death, and according to the Biblical account they continue to be visible even after his resurrection (John 20:27). We are told bluntly that the body of this resurrected God is a maimed one. This fact of a disabled God, Eiesland said, should fundamentally unsettle our cultural ideas about aspiring to a perfect body, and what it means to have an imperfect one.

In *The Body of God*, the feminist theologian Sallie McFague[21] argues that the incarnation (leaving aside for the moment whether you consider this to be fact or metaphor) says something important about a divine willingness to participate in embodiment. This claim stands in stark contrast to the deep unease about the body that runs through so much of Christian history. McFague doesn't directly discuss disability

---

[19] Hans S. Reinders, *Receiving the Gift of Friendship: Profound Disability, Theological Anthropology, and Ethics*. Wm B Eerdmans, 2008.
[20] Nancy L. Eiesland, *The Disabled God: Toward a Liberatory Theology of Disability*. Abingdon, 1994.
[21] Sallie McFague, *The Body of God: An Ecological Theology*. Augsburg, 1993.

in her highly influential book, but when she states that 'the body of God is not *a* body [my emphasis], but all the different, peculiar, particular bodies about us,' it must surely make us reflect on our understanding both of the metaphor of the body of God, and of human diversity – including disability. What does it mean if the body of God, the material manifestation of what Quakers call the Inward Light, always includes disabled bodies, or (as in Eiesland's book) is disabled itself?

Several contemporary theologians have moved away from a primary focus on the meaning of an individual disabled body, to a concern with what a disability perspective can offer to our communities, faith groups, and society as a whole. Earlier, I proposed an emerging dissident interpretation of the healing narratives, where the point of those stories lies not in the magical removal of an impairment but in the re-establishment of the relationship between the community and those who have been excluded from it because of their disability. If that interpretation makes sense, then it implies that inclusion – a gathering back into the human community – has a strong spiritual mandate. What implications does this have for how our own communities, including the Religious Society of Friends, can make connection and relationship a reality?

Deborah Beth Creamer is another theologian who has written extensively on disability and has developed what she calls a **limits** model.[22] She is sharply critical of the prevalent, and usually unchallenged, belief that to be limited in any way is always a negative thing. The corollary of this belief is that the fewer constraints a person experiences, the better; and from that it is a very short step to the conclusion that the fewer limits one has, the closer one is to perfection – or, put in religious language, the more like a perfect God one can be. For those of us of a philosophical bent, this is a manifestation of a debate that has gone on for millennia (and that I can't pursue further here): the question of what, exactly, perfection means, especially if one is trying to imagine perfect divinity. In this context, it's worth noting the longstanding tradition that says God can be infinite and unlimited,

---

[22] Deborah Beth Creamer, *Disability and Christian Theology: Embodied Limits and Constructive Possibilities*. Oxford University Press, 2008.

and at the same time deliberately self-limiting to open up a space of freedom and choice within creation.[23]

In a similar vein, Creamer wants to acknowledge that physical and mental limits are integral and unavoidable aspects of the human condition. Limits are not solely about closing off possibilities; they can also be productive and generative. This isn't quite as simplistic as saying that a disabled person is forced to be extraordinarily creative in devising ways to overcome their physical or mental limits, although that can be true. It is more fundamentally about the *conditions of possibility* for creativity. Any artist will tell you that the constraints inherent in the medium of creation are essential to the creative process. Watercolours and oil paints produce very different pieces of work, and engage very different forms of creativity, because of the distinctive properties of the two media. Ironically, a total absence of limits or constraints means a total lack of properties – in other words, a void, a state of sterility and death, not creativity.

Interestingly, Creamer also has reservations about the language of 'disability pride' used by many activists, because she sees it as still trapped within the idea of difference as defect. Disability pride says that having an impairment or disability should not be equated with being limited – which is fine, but still isn't dissident enough to challenge the basic belief that limits are necessarily obstacles that keep us from reaching our full potential. Limits are more than just universal and ubiquitous; they are *necessary* to have any meaningful personal or community identity at all.

This is not to claim that all limitations are benign, and it certainly isn't a proposal that we should never try to overcome them. Serenely accepting all limits to human experience would mean no social or medical progress, resigning people to (among other things) unbearable pain and early death, and gross social injustices in the form of the artificial boundaries placed around what people can do or be. But dissidently inverting the conventional logic about the badness of limits

---

[23] The idea of a voluntarily self-limiting God is most closely associated with the 20th century German theologian Jürgen Moltmann, particularly in his *The Crucified God*, first published in English in 1973. Moltmann died at the age of 98 as I was correcting the proofs of this lecture.

can bring into clearer view some of our assumptions about the ranking of normality and difference. These assumptions need to be subjected to thoughtful ethical and spiritual examination by each of us as well as by society collectively. Why such reflection is urgently needed right now is the subject of the next section.

Disability in Reproductive Technologies

The everyday situation of people with disability today is a complex mosaic of better and worse. On the plus side, there is more acknowledgement than there once was that disability is common. According to the World Health Organization, around 1.3 billion people globally experience a significant level of disability,[24] while the Australian Institute of Health and Welfare gives a figure of around 4.4 million people, or 18% of the population, for Australia.[25] Over the last half century, driven by the activism of disabled people and their families, there have been genuine advances in equality for and inclusion of disabled people. The United Nations Convention on the Rights of Persons with Disabilities, which opened for signature by states in 2007 and came into force in Australia in 2008, is emblematic of this, but was only possible after decades of working for incremental change in attitudes towards disability.

But although their human rights are better acknowledged than ever before, people with disability are still excluded and discriminated against at all levels of society, a phenomenon often described as *ableism*. In 2011, the *World Report on Disability* found that everywhere in the world, disabled people are more likely to be poor, unemployed, lack access to education or healthcare, and suffer most harm during disasters or conflicts, and every report since then confirms that the situation has not improved.[26, 27] It seems that the lingering reluctance

---

[24] World Health Organization, Fact Sheet on Disability, 2023: https://www.who.int/news-room/fact-sheets/detail/disability-and-health

[25] Australian Institute of Health and Welfare, Australian Government, People with disability in Australia, accessed 19 May 2024. https://www.aihw.gov.au/reports/disability/people-with-disability-in-australia/contents/people-with-disability/prevalence-of-disability

[26] World Health Organization, *World Report on Disability*, 2011. https://www.who.int/publications/i/item/9789241564182

[27] United Nations Office for Disaster Risk Reduction, *Global Survey on Persons with Disabilities and Disasters*, 2023. https://www.undrr.org/report/2023-gobal-survey-report-on-persons-with-disabilities-and-disasters

to encounter anything perceived as abnormal still presents a barrier to genuine inclusion.

It is in this ambivalent context that biomedical science is developing capabilities that, for the first time in human history, can reliably identify (some) impairments long before birth. **Prenatal testing and screening** for a range of anomalies has become a routine part of antenatal care in wealthier countries. Prenatal screening for Down syndrome has been available in Australia since the late 1960s, along with ultrasound to detect other structural variations. Advances in genetic science are now opening the way to identifying a rapidly expanding number of variations in the genetic make-up (the genome) of the developing baby. Many of these variations have no noticeable effect on the baby's life, while others will lead to disabling conditions of varying severity.

In countries like Australia, prenatal screening programmes are offered with the aim of providing prospective parents with more information about their future child, and therefore with more options. In practice, in most cases the choice is between continuing with the pregnancy or terminating it: at the moment, there are only a relatively small number of conditions for which having that foreknowledge means a treatment can be offered.

Decisions about termination of pregnancy that can follow prenatal screening are ethically and emotionally fraught. Recently, a process called **preimplantation genetic diagnosis, PGD,** has become available, in which an embryo is created by in vitro fertilisation (IVF) and its genes are then tested before deciding whether or not to attempt a pregnancy with it. PGD is considered by some people to be morally and emotionally easier than testing and possibly aborting a foetus, but it is still a decision over *the kind of life* that is coming into being. And, on the horizon, though not yet practised in any country and overtly forbidden in many,[28] is the prospect of directly manipulating the

---

[28] In 2018, an attempt to manipulate the genes of two children prenatally was reported by a Chinese scientist, He Jiankui. Although there has been little further information, there is no evidence that the attempt was successful. It is still very unclear to what extent the Chinese government knew of and supported his work, but, following the strongly negative global response, he appears to have been detained by the authorities and went on to serve three years in prison. He was released in April 2022.

genes of an embryo at a very early stage of development to prevent a disabling condition. Such **gene editing** presents ethical challenges that are different again, not about ending the life of a foetus or embryo but about changing the identity of one that is already here.

I've gone into this in some detail to highlight a few things. First, many people clearly *do* want the opportunity to screen or test for disabling conditions. Research shows that most women offered prenatal screening take it up. The desire to test is not always with the aim of terminating the pregnancy if an anomaly is found. Sometimes, the prospective parents just want the knowledge to prepare themselves, mentally and practically, for a different kind of future than the one they had envisaged, with a different kind of child.

Nevertheless, it's also true that when a serious anomaly is detected in pregnancy, in most cases the parents opt for termination. The concern for bioethicists like me is not about the legitimacy of abortion per se, but unease about the routinisation of prenatal testing and the risk that the capacity to control who comes into the world might reduce the (already fairly low) societal tolerance of human variation. Routinisation combined with an expectation of perfectibility can lead to a lack of thought about whether, in any individual case, the right choices are being made. Prenatal selection is offered as a means for parents to exercise more choice and control, and, in modern society, having choice is generally considered an unalloyed good. Yet there is substantial evidence to suggest that, to prospective parents, the whole process of prenatal screening can feel more like doing whatever is expected by society or the medical profession (women who have been through this often say 'it's like being on a conveyor belt') than having more choice.

That there *are* strong societal expectations about the kind of reproductive choices people should make is demonstrated by the thorny issue of so-called 'choosing disability'. A very small number of people with what are considered disabling conditions have expressed the wish to select *for*, rather than against, children with those disabilities. It is not common, and it almost always involves members of the Deaf community[29] who consider themselves to be a minority cultural group rather than disabled, and who want their child to share their distinctive cultural feature. Now, this is clearly an exercise of reproductive choice, yet it has invariably been met with disapproval and, in some cases, legal steps to prevent it happening.[30] Note that I'm not arguing here that opting for rather than against disability is necessarily morally ok, only that the reactions make it abundantly clear that offering selective reproductive technologies has a more complex motivation than just the maximisation of parental choice. It's about enabling people to make societally approved choices.

Many disability activists and scholars feel strongly that any form of reproductive selection to avoid disability is an expression of hostility to disabled people and, as such, runs counter to the goal of increasing inclusion and the acceptance of difference. They argue that prenatal selection to prevent disability is eugenic, just the latest step in the long history of ideas about improving the 'quality' of the population by getting rid of disabled people. The eugenic eradication of people with disability and some mental illnesses reached its nadir during the Nazi Third Reich in the mid-20th century.[31] But others, including myself, believe that selection *can* be an appropriate response to some disabling conditions that have so severe an impact only a sadist would remove the option of avoiding a life with such a level of suffering, for individuals and their families. I think that what is 'eugenic' here is not the technology itself, but the desire for easy, or at least unambiguous, choices.

Through most of human history, our beliefs about the meaning of disability were manifested in our attitudes and behaviour towards real, living disabled people. Being able to enact those beliefs by controlling what kind of people enter into life and become part of our world is a very different responsibility. This is why I suggest that we are at a

---

[29] By convention, deafness as audiological hearing loss is denoted by a lower case 'd', Deafness as a cultural identity by an upper case 'D'.

[30] I wrote a short chapter on this, entitled 'Being Disabled and Contemplating Disabled Children', in *The Disability Bioethics Reader*, edited by Joel Michael Reynolds and Christine Wieseler, published by Routledge in 2022.

[31] See Michael Robertson, Astrid Ley, Edwina Light, *The First into the Dark: The Nazi Persecution of the Disabled*. UTS Press, 2019.

crucial point in the human encounter with difference, disability and normality. The biomedical technologies that enable us to identify and select for or against difference are themselves the result of our (I would say, God-given) drive to search constantly for ways to make people's lives better, but they hand us enormous power that must be used carefully. As a society, how can we best take on this responsibility with confidence that we are making the right decisions?

One very basic requirement would be familiarity with the experiential realities of living with a disability. Making morally endorsable judgements about quality of life, or a life in which someone can flourish, simply isn't possible without a well-grounded knowledge of real disabled lives. By paying closer attention to personal experiences of disability we would get deeper insight into lives that are often overlooked or ignored, lives that are different but, often, no worse than anyone else's.

But for personal experience to be more than just a set of anecdotes or free-floating claims about truth, it needs to be reflected on, individually and collectively. The goal is to distinguish those situations in which bodily difference really is just a neutral variation, from those where it is an impairment; or situations where impairment is unavoidably associated with suffering, versus others where suffering could be reduced by re-making the world around greater accessibility and acceptance. This process of discernment is a familiar part of Quaker faith and practice, and the subject of the next section.

Disability and Discernment

The question for Friends today is how we are called to respond to disability and impairment, and difference in general, as part of building a world in the 21ˢᵗ century and beyond that aligns with our testimonies, those 'outward and visible signs' of our inward commitment to the Spirit. In this part, I highlight three themes of central importance to disability:

- being embodied;
- being creatures that are vulnerable and dependent, and not in total control of what happens to us; and
- being a just and inclusive community.

I consider how, when viewed through the dissident lens of Quaker Testimony,[32] these themes illuminate our understanding of disability, difference and normality.

Being Embodied

All of us, disabled and nondisabled alike, are embodied creatures. In fact, Christianity is a religion for which the body is absolutely central: the doctrine of the incarnation makes the distinctive faith claim that divinity can, at one and the same time, utterly transcend the material world and also be comfortable walking around in a bag of skin. Despite this, Christianity is also well known for its history of profound ambivalence towards the facts of the human body, and particularly to bodies that deviate from the norm.[33] Earlier, I mentioned the emergence of 'body theology' and how it places embodiment at the centre of its approach. Yet even here, a closer look suggests that body theology focuses almost exclusively on issues of sexuality, with relatively little attention paid to bodily *difference* except the differences of sex, gender, and sexual orientation. Several disability commentators have noticed that body theology on the whole continues to take it for

---

[32] I'm using the convention that the general concept of Quaker Testimony is indicated by an upper case 'T', individual testimonies by a lower case 't'.

[33] For example, by not being male or white.

granted that the ideal body is a nondisabled one: 'Even cutting-edge work on gender, race, and sexuality still assumes a generically healthy body to be normative.'[34]

My earlier discussion could only gesture towards the rich resources now available to us with which we could develop a more open and generous relationship towards variant bodies. When the 'normal' body is taken as the starting point for all thinking about disability, the disabled body is defined in terms of its deficits compared to the nondisabled norm. In this perspective, disability is always a matter of falling short. Instead, we could acknowledge that, if bodily difference is inevitable and being limited by our bodies a universal experience, then a binary system of classification in which disabled people are split off from the nondisabled just doesn't work. In the everyday world, of course, binary systems have many advantages, particularly in what might be called social bureaucracy. For you to receive a disability benefit from the NDIS, the State has to agree that you belong to the group of people eligible to get one. A line has to be drawn; you are either in or out, so you do or don't get support.

What I've called dissident thinking, however, calls us to step outside this system altogether. **Rather than start from a binary categorisation,** trying to decide where to place the line between normal and disabled and developing ever more refined criteria to identify who falls into either category, we could consider how any person's set of embodied limits play out in the context in which they live, and how they might variously hamper or enable their chances of living a flourishing life. The gap that needs to be filled here is in our knowledge of living with illness and disability, a gap that is too often filled with guesswork and presumptions of misery. Yet the more we know about the lives of people with disability, the clearer the message that they can be so much more fluid, nuanced, multifaceted, subversive, *extravagant* than a single simple label suggests.

Here, though, I want to add a major note of caution. Rejecting the binary classification of disabled/nondisabled is one thing; ricocheting

---

[34] Deborah Beth Creamer, *Disability and Christian Theology: Embodied Limits and Constructive Possibilities*. Oxford University Press, 2008, p. 55.

to the other extreme and relabelling all disability as neutral variation is quite another. Disability advocates have sometimes suggested that, to avoid being eugenic, it is necessary to welcome all forms of bodily difference without reservation, because all differences are equally valuable parts of our human diversity (the 'let a hundred flowers blossom' model). But I think that, fundamentally, this response is the same as the ableist demand that all impairment should be eradicated. In both cases, there is a deep disinclination to engage in any meaningful way with the actual reality of living with disability.

At the start of this lecture, I noted that disability is often used as if equivalent to other equity cohorts (sex, gender, race, sexual orientation and so on). In reality, it is not the same as any of these, because some forms of disability come with a level of intrinsic disadvantage that can't be made better simply by removing discriminatory attitudes or disabling barriers. Some examples might make this clearer. It is likely that the disadvantages of being black in a majority white society would disappear if people weren't racist; the disadvantages of being a woman in a patriarchal society would vanish if there were full sex and gender equality. It's not that the material differences of skin colour or sex would disappear, but that the differences would no longer be linked to disadvantage or suffering.

For at least some of the things we identify as impairments or disabilities, though, this isn't true. Social-relational models of disability predict that many apparently severe disabilities would be much less problematic if society could be made more inclusive, and some might even become neutral differences rather than impairments. Nevertheless, there would remain conditions for which no amount of changed attitudes or accessible buildings could eradicate all the disadvantages. In a more disability-friendly world a person with the genetic condition osteogenesis imperfecta (brittle bones) might be much more able to access education, travel to work, have a social life and so on, and these would all be improvements, but the fractures, pain, fragility and other medical consequences would still be there. This is what makes it impossible to say sweepingly, 'I welcome the diversity that is disability' in *quite the same way* that we would say 'I welcome diversity in race' or 'in gender'.

The Quaker response to diversity is rooted in the testimony of equality. The testimony doesn't concern itself with particular characteristics or abilities, but with the recognition that 'the Spirit of God dwells in every person'.[35] Calls for equality are often mistaken for the claim that everyone is the same, or even that everyone should be treated the same. But to practice true equality is to be able to hold in mind, at one and the same time, that which is common to all of us *and* the particularities that are unique to individuals and that shape their distinctive strengths, capabilities, needs and vulnerabilities. Equality doesn't mean ignoring genuine differences, through a misplaced sense of kindness or a fear of conflict. It means standing on a bedrock of knowing that the Inward Light, like love, is not measured out differently for different people: we are all equal in the sight of God. This knowledge is the ground of our commonality as people, and it gives us the security from which to engage fully with embodied difference.

Being Vulnerable, Dependent, and Not in Control

For much of modern western history, the defining feature of the people who are taken seriously and valued as members of society has been *autonomy*: the capacity to be an independent agent, determining for themselves how they want their lives to go and being able to take the necessary steps to make that happen. People who are in any way obviously dependent on something or someone outside themselves, including people with disabling conditions, have been and often still are seen as lacking full autonomy. In this view, a wheelchair user is dependent on the presence of lifts and ramps, or electricity for a power chair; someone with a hearing impairment may be dependent on having a hearing aid and batteries, or on the ability of those around them to use sign language; a diabetic requires regular insulin injections to survive. The unspoken assumption that the more dependent someone is, the less of a real person they are, continues to be a significant barrier to disabled people's equality.

---

[35] Australia Yearly Meeting, *This We Can Say: Australian Quaker Life, Faith and Thought,* 2003, 3.52.

Independence is a concept that is closely interwoven with others such as vulnerability, need, and care. The logic goes like this: because of their impairments, people with disability have special vulnerabilities that make them more dependent (wheelchair users are dependent on lifts); conversely, being dependent on something is itself a vulnerability (wheelchair users suddenly become more vulnerable if the lifts break down). A dependence is also an unusual need, and unusual needs are seen as requiring extraordinary forms of care. For many, the defining feature of disabled people is not their variant body form or function, but that they need direct or indirect care provided by others. The presumption then is that this presents extraordinary burdens for families or the State, especially when the need for care is equated with the failure to be a productive (i.e. wage-earning) member of society.[36]

This neat model ignores the fact that even able-bodied, healthy adults all live within elaborate webs of dependency and care. Nor is this confined to a few stages of the life course or unusual circumstances. We recognise that infants, very old people and those who are sick are highly dependent but choose not to acknowledge that few adults would survive if they had to provide for themselves absolutely everything a human organism needs. The most independent person will nevertheless need food, which means somewhere (like a shop) where they can obtain it, and therefore food producers, supply chains, transport, fuel, an economy. We live in groups because as individuals, we need the activities of others to live at all. These needs are so universal they are not even seen as dependencies, until a crisis of some form erupts, and individuals then discover what happens when those needs are not met. In reality, the much vaunted ideal of total independence and invulnerability is a delusion.

The theologian Kathy Black has suggested that disability offers an insight into the real interdependence of everyone, disabled or not. The dependencies of disabled people only seem unusual because the needs for sign language interpretation, a ramp, or somewhere to have a rest are highlighted as anomalous, while needing a car to get to work

---

[36] Which links to another set of challengeable assumptions, too big to go into here, that only people who are economically productive can be considered full and valuable members of society.

or someone to come and fix your pool is not. And Black names the societal refusal to recognise the fact of universal interdependence as a sin.[37]

It is interesting that, while our culture overvalues independence, we also seem deeply ambivalent about the notion of vulnerability. Much of my research over the past 25 years has explored people's attitudes towards disability and difference in various biomedical settings. I don't remember a single one out of hundreds of research participants who thought that being perfectly *in*vulnerable is desirable; in fact, some of them described invulnerability as fundamentally inhuman.

The intuition that it's necessary to retain some level of vulnerability goes hand in hand with the sense that being in control of everything, in other words being able to decide exactly the course you want your life to take, isn't always a desirable thing. It is challenging to articulate a rationale for this, especially when faced with the evidence that controlling things like the impact of disease has made life much better for most people. It's hard to argue against the benefit of eradicating smallpox. Very few bioethicists have seriously addressed what the American scholar Erik Parens describes as 'the goodness of fragility'[38]: the sense that being vulnerable to events or lacking total control over them can be good in itself. Yet this intuition is strong and persistent. While many people have told me how much they value the foreknowledge that prenatal screening gives them, others would rather not know, saying they're glad they had their children before prenatal testing was widely available because for them the decision would have been impossibly difficult. They can't imagine life without the presence in the world of the child they now love and value.

With sporadic exceptions, like the Christmas carols that sentimentalise the fragile baby Jesus, our religious culture is equally addicted to invulnerability and the goal of achieving control. The God of the church tradition is both perfect and all-powerful. If, as we are

---
[37] Kathy Black, *A Healing Homiletic: Preaching and Disability*. Abingdon Press, 1996.
[38] Erik Parens, The goodness of fragility: on the prospect of genetic technologies aimed at the enhancement of human capacities. *Kennedy Inst Ethics J* 1995. Jun;5(2):141–153. doi: 10.1353/ken.0.0149

told, people are created in the image of God,[39] then linking divinity with perfection, absolute power, and control inevitably leads to the conclusion that imperfect, vulnerable, dependent people – including people with disability – can only ever be a *substandard* image. But if we take 'that of God in everyone' seriously then being disabled cannot be synonymous with being a depleted version of a person, or marginal to the life of the community. There are some important questions here for society as a whole about whether increasing control over every area of life is necessarily something we want. For Quakers, I think it warrants particular reflection in view of our testimony of simplicity. It might be that the testimony calls us to take a dissident stance, by consciously relinquishing at least some of our attempts to control the kinds of bodies people have, or the people who come to occupy the world. Clearly, we can't lurch to the opposite extreme of radical powerlessness (there's no way I'm going to give up attempting to control my body's rejection of my donor liver, for example). So how do we find that Goldilocks point of just enough vulnerability but not too much?

Thinking dissidently about vulnerability and dependence offers an alternative insight into their importance for human life and our relationship with God as well as with each other. Of course, none of this should suggest that disabled people exist to serve as a kind of audio-visual aid, reminding the rest of us on occasion about our own fundamental vulnerability. It is more about opening up to a different view of humanity, in which 'the works of God' are shown through bodies in relation to other bodies. Our likeness or closeness to divinity (or at least one aspect of it) just *is* our ability to be in relationship with others.[40] It isn't that particular people more closely reflect the Spirit in their dependence and their need for care, or on the other hand in their capacity to provide care for others: either would be uncomfortably close to what might be called the 'least of these' model, which has too often been used to marginalise and infantilise people with disability. Instead,

---

[39] Entire theological careers have been built on debating exactly what that means, so I'm not going to enter into that discussion here.

[40] I want to offer some reassurance to those readers who, like me, are strong introverts, that being in relationship with others doesn't mean interacting with other people all the time. The concept of relationality I have in mind is more like a network, with many different strands of linkage, some more distanced than others.

I'm suggesting that the Spirit is constituted within those networks of care and dependency: neither by those in need nor by those providing care, but in what goes on between them.

Being in Community

The word community is used rather loosely today to mean any (minority) grouping: think of 'the Black community', 'the gay community', even 'the disability community'. I say minority, because interestingly it's rarely used for the dominant set of people -- we don't speak of 'the white community', for example. Although it can be a convenient shorthand for 'all those people sharing this characteristic', at worst it can be a covert way of othering an unfamiliar group.

Behind the vision of self-sufficiency that our culture values so highly lies the notion that a person exists as a discrete individual before becoming part of the relationships that entwine her in the social world, and that being in community is optional. Quakers ought to have a more nuanced understanding of what being a community really entails. We started out as a bunch of renegades, under threat from the political and religious establishment and heavily dependent on each other practically, emotionally and spiritually. What Nancy Eiesland writes here is a powerful rebuttal: the metaphor of the disabled God shows a 'God for whom interdependence is not a possibility to be willed from a position of power, but a necessary condition for life.'[41] We are in community, or we don't exist at all.

This also has something to say about the testimony of sustainability – not among the earliest Quaker testimonies, but one that has grown in importance along with our realisation of humanity's destructive impact on this earth. Sustainability is about the continuation of life, in all its diversity. Diversity tends to be understood first and foremost in terms of ways of living. In a pluralist society, it means being enriched by the huge range of lifestyles, ethnicities, religions, behaviours, and interests of its citizens. By contrast, bodily diversity, as we've seen, tends

---

[41] Nancy L. Eiesland, *The Disabled God: Toward a Liberatory Theology of Disability*. Abingdon, 1994, p. 103.

to be overlooked, or even rejected as a possible good. To a biologist,[42] however, diversity is not simply a nice thing to have, but a necessity for species survival. Any organism's genetic make-up is very similar to, but also very different from, all other organisms of the same species. In evolutionary terms, bodies evolve to suit particular environments, and this occurs through changes in the genes. Random genetic variations are thrown up by chance in every generation, and usually make no difference to the individual. Sometimes, though, a genetic variation makes an organism a little better suited to its environment, and then the variation is likely to be passed on to subsequent generations. Sometimes, it results in a change that is disabling, and so it may or may not persist depending on just how much of a disadvantage it is. In any case a population without genetic diversity is in deep trouble, because it has no means of adapting if and when circumstances change. The price of this reservoir of adaptability, however, is genetic variation that is damaging.

A flourishing, sustainable community has to vary, biologically and socially. To put it in theological terms, God's purpose is achieved not through reaching some kind of ideal state or form, but through the presence of 'the energy for unleashing multiple forms of corporeal flourishing'[43] in the constant flux of creation.

The real test of community is how we live with those others whose lives appear to be radically different from our own. In the past, the responses of religious communities – not just within the Christian tradition – to anyone seen as *other* have often been frankly disappointing. Recent writing by disability scholars documents the many failures of the church community's engagement with disabled people, ranging from the embarrassing to the deeply traumatising. Even today, people with disability are still told that they could be healed if only their faith were great enough (with the implication that those who are still disabled must be lacking in faith); conversely, that their illness or disability is proof that God has specially selected them

---

[42] Or former one; although I think that you can take the girl out of science, but you can't take science out of the girl.

[43] Sharon V. Betcher, *Spirit and the Politics of Disablement*. Fortress Press, 2007, p. 52.

because of their greater than usual strength of faith; that they provide an opportunity for faith group members to show compassion; that their experience gives them a unique spiritual wisdom; or, failing all else, that they will be compensated for their suffering in the world to come. None of these, I suggest, is likely to make disabled Friends feel like equal, valued and welcome members of the Meeting.

Communities demonstrate whether someone belongs or not through their behaviour towards them. Difference and diversity, including disability, may be overtly and loudly celebrated, while at the same time disabled people find themselves excluded if the modifications they need for access are not provided. I used to know someone in the UK who was a lifelong wheelchair user and attended a Quaker meeting for several years. Every single week, on entering the lobby of the meeting house, he had to push aside chairs and tables that were placed in such a way that they blocked his path to the room where meeting for worship was held. Every single week, he was observed doing this, and the following week would be greeted with exactly the same obstacle course. Eventually he stopped coming to Meeting at all, partly out of exasperation, but also disillusioned by the hypocrisy. Claiming to recognise that of God in everyone sounds lovely but is meaningless if no one makes much effort to ensure that God, in the form of a disabled person, can actually get into the building.

Thinking about the place of people with disability within the Quaker community brings me to a final point: the fact that communities share a mutually comprehensible language. Everyday speech is dynamic, and regularly scrutinised for terms that were once acceptable but are now considered misleading, derogatory or offensive, particularly to minority or marginalised groups, to be replaced with 'cleaner' language. (Almost inevitably, though, the new word becomes contaminated by association over time and has to be replaced by something else in its turn.) The disability world is full of examples. 'Down syndrome' has replaced 'mongolism', 'cripple' is no longer used, 'deaf and dumb' has thankfully been laid to rest. These changes show greater respect to the opinions of the people most affected and are to be welcomed.

But there is also a trend towards using unhelpful and sometimes patronising euphemisms for disability. Phrases like 'differently abled', 'challenged', or 'special needs', often put forward as more progressive alternatives, are nearly always devised by nondisabled people who feel awkward around disability. 'People living with disability' or 'with lived experience of disability' are currently popular in Australia but infuriate many actual disabled people. I've encountered the use of 'people with lived experience' (dropping the d-word altogether) as a coded way of referring to disability, even though logically it refers to everyone since nobody alive has dead experience. More seriously, euphemism at this level also sends out a clear message that disability is too shameful to speak about.

Words can seem trivial in comparison to action, but how we talk about disability is spiritually important because what we say is as much a part of Testimony as how we act. Straightforwardness in speech is particularly relevant to two testimonies: simplicity and integrity. We generally think about the testimony of simplicity in the context of lifestyle, such as reducing consumption and living sustainably, but it also has implications for other areas of life. At the start of this lecture, I pointed out that our everyday language makes it hard to talk in a clear and value-free way about disability, and this isn't helped by adopting euphemisms. Needless to say, plain speaking doesn't mean being tactless or brutal, but it does mean avoiding language that obscures more than it clarifies. (Academics like me are particularly prone to this.)

And integrity? As much as being honest or keeping our word, integrity means being able to stand by what we say and do. That in turn involves a lot of prior hard work: of thinking, listening, and above all discernment. The final section of this lecture reflects on this.

A Dissident Stance on Disability and Difference

Astute readers will have noticed that this lecture is scattered with questions to which, disappointingly perhaps, I don't give answers. One reason for this is that, although bioethicists are often expected to provide definitive guidance on morally troubling medical issues, in practice we aren't particularly good at it. Our skills lie more in analysis, asking question after question about what is going on in a situation of moral trouble, teasing apart the various threads in an ethical tangle to see where the real problem lies. As an empirical ethicist, I've always been less interested in theory than with bringing my analytic tools to bear on real life situations and how affected people respond to them, because I believe that this will bring us closer to a more complete understanding and, ultimately, lead the way to better solutions.

A second reason is that, when it comes to a matter like disability that affects so many people – potentially, everyone – it just isn't appropriate for a single bioethicist to have a bit of a think and then tell everyone what to do. There needs to be deep and thoughtful discussion across every part of every community, each bringing their own experiences and perspectives. Within the Religious Society of Friends, our focus will usually be on seeking the Spirit's guidance on how to embed justice and equality for disabled people inside and beyond our own community and in a rapidly changing context. In an earlier part of this lecture, I described the emerging biomedical technologies that effectively give us power to control whether people with disabling genetic conditions are born. It's important to emphasise that even if prenatal screening were to result in no babies with genetic conditions coming into the world at all there would still be plenty of disabled people around, because the vast majority of disability happens *after* birth as a result of ageing, accident or illness. I've concentrated on techniques of genetic selection because these are areas I know best, and because the power they offer is so new it should make us pause, take a deep breath, and consider what we *really* think about disability and difference. It should not be forgotten, though, that there are many aspects of life, other than prenatal selection, where people with disability experience injustice, exclusion and disadvantage, and which should prompt equal concern.

I've suggested that to engage better with disability we need to think not just differently but *dissidently*, as Quakers have historically shown themselves more than able to do. A dissident approach should challenge some things that have long been taken for granted in the discussion, such as that disability can easily be separated from bodily difference more generally, or that there is an objective standard of normality. Dissident thinking includes being 'open to new light, from whatever source it may come',[44] but that doesn't mean accepting it uncritically – whether it's scientific research, spiritual insight, the experience of disabled people, or the random thoughts of a Quaker bioethicist. Dissidence also means being prepared, if necessary, to stand aside from the phrases, models or entire ideologies currently in fashion if they prove unhelpful in guiding us towards better understanding. Note that dissidence isn't being contrary for the sake of it, but the effort to abandon the reflexes of thought and belief that get in the way of true discernment.

Discernment can be an individual process, but I believe it is generally best done in community. True community is essentially about balancing interdependence and difference, and isn't always easy or comfortable, something else that Friends will readily recognise. In 1667, Isaac Penington wrote that 'Our life is love, and peace, and tenderness; and bearing one with another, and forgiving one another, and not laying accusations one against another; but praying one for another and helping one another up with a tender hand'[45], and the fact that he felt it needed saying suggests that the Quaker community at the time may have needed reminding. It tells us that, in building a community and a world that is truly inclusive of difference *and* as fair as possible to all its members' rights and needs, we shouldn't expect to find immediate unity or clarity. People with disability themselves don't

---

[44] Britain Yearly Meeting, *Quaker Faith and Practice: Advices and Queries*. London, 1995.

[45] Isaac Penington, writing to Friends in Amersham in 1667, quoted in Australia Yearly Meeting, *This We Can Say: Australian Quaker Life, Faith and Thought*, 2003, 3.58.

share a consensus view on difficult issues like prenatal selection, or voluntary assisted dying, or even the spiritual meaning of vulnerability and fragility. And, fundamentally, disability and difference *just are* complex and nuanced phenomena. Our only hope for properly understanding them is to include multiple perspectives, not just the ones we happen to agree with.

I will end with a question (or perhaps a Query): How can we live generously and adventurously with embodied difference, so that disabled and nondisabled people alike can flourish in our various communities?

Further Resources

Here are some resources for further reading, for those interested.

Sources referred to in the Lecture

Australia Yearly Meeting. *This We Can Say: Australian Quaker Life, Faith and Thought*, 2003.

Betcher, Sharon V. *Spirit and the Politics of Disablement.* Fortress Press, 2007.

Black, Kathy. *A Healing Homiletic: Preaching and Disability.* Abingdon Press, 1996.

Block, Jennie Weiss. *Copious Hosting: A Theology of Access for People with Disabilities.* Continuum, 2002.

Borsay, Anne. *Disability and Social Policy in Britain since 1750: A History of Exclusion.* Palgrave, 2005.

Britain Yearly Meeting. *Quaker Faith and Practice: Advices and Queries.* London, 1995.

Creamer, Deborah Beth. *Disability and Christian Theology: Embodied Limits and Constructive Possibilities.* Oxford University Press, 2008.

Eiesland, Nancy L. *The Disabled God: Toward a Liberatory Theology of Disability.* Abingdon, 1994.

Hick, John. *Evil and the God of Love.* Palgrave, 2010 (1972).

Kittay, Eva Feder. *Learning From My Daughter: The Value and Care of Disabled Minds.* Oxford University Press, 2019.

McFague, Sallie. *The Body of God: An Ecological Theology.* Augsburg, 1993.

Moltmann, Jürgen. *The Crucified God.* 40th anniversary edition. Fortress, 2015.

Parens, Erik. *The goodness of fragility: on the prospect of genetic technologies aimed at the enhancement of human capacities. Kennedy Inst Ethics J* 1995 Jun;5(2):141–153. doi: 10.1353/ken.0.0149

Reinders, Hans S. *Receiving the Gift of Friendship: Profound Disability, Theological Anthropology, and Ethics.* Wm B Eerdmans, 2008.

Robertson, Michael, Ley, Astrid, Light, Edwina. *The First into the Dark: The Nazi Persecution of the Disabled.* UTS Press, 2019

Scully, Jackie Leach. *Disability Bioethics: Moral Bodies, Moral Difference.* Rowman and Littlefield, 2008

Scully, Jackie Leach. *Being Disabled and Contemplating Disabled Children.* In: Reynolds, Joel Michael and Wieseler, Christine. *The Disability Bioethics Reader.* Routledge, 2022, p 116-124

Shakespeare, Tom. *Disability Rights and Wrongs Revisited.* Routledge, 2014.

Further Reading

Some additional resources on faith and/or disability.

Loukin, Esther. *How We Can Change Society: the 2023 Swarthmore Lecture.* Britain Yearly Meeting, 2023.

The 2023 Swarthmore Lecture to the Religious Society of Friends in the UK also, coincidentally, looked at disability.

Melcher, Sarah J, Parsons, Mikeal C, Yong, Amos (eds). *The Bible and Disability: A Commentary.* Baylor University Press, 2017.

Parens, Erik, Asch, Adrienne (eds). *Prenatal Testing and Disability Rights.* Georgetown University Press, 2000.

Reynolds, Thomas E. Vulnerable Communion: *A Theology of Disability and Hospitality.* Brazos Press, 2008.

Segal, Julia. *The Trouble with Illness: How Illness and Disability Affect Relationships.* Jessica Kingsley Publishers, 2017.

Some Personal Accounts

Of the many personal accounts of disability available, here is just a handful, very different in style and content, that I've read and enjoyed.

Beck, Martha. *Expecting Adam: a True Story of Birth, Rebirth, and Everyday Magic.* Random House, 1999.

Findlay, Carly (ed). *Growing Up Disabled in Australia.* Black Inc, 2021.

Murphy, Fiona. *The Shape of Sound.* The Text Publishing Company, 2021.

Spufford, Margaret. *Celebration.* Fount, 1989.

www.ingramcontent.com/pod-product-compliance
Lightning Source LLC
Chambersburg PA
CBHW070104100426
42743CB00012B/2650